MW01065942

Put Beginning Readers on the Right Track with ALL ABOARD READING™

The All Aboard Reading series is especially for beginning readers. Written by noted authors and illustrated in full color, these are books that children really and truly *want* to read—books to excite their imagination, tickle their funny bone, expand their interests, and support their feelings. With four different reading levels, All Aboard Reading lets you choose which books are most appropriate for your children and their growing abilities.

Picture Readers—for Ages 3 to 6
Picture Readers have super-simple texts, with many nouns appearing as rebus pictures. At the end of each book are 24 flash cards—on one side is the rebus picture; on the other side is the written-out word.

Level 1—for Preschool through First-Grade Children
Level 1 books have very few lines per page, very large type, easy words, lots of repetition, and pictures with visual "cues" to help children figure out the words on the page.

Level 2—for First-Grade to Third-Grade Children
Level 2 books are printed in slightly smaller type than Level 1 books. The stories are more complex, but there is still lots of repetition in the text, and many pictures. The sentences are quite simple and are broken up into short lines to make reading easier.

Level 3—for Second-Grade through Third-Grade Children
Level 3 books have considerably longer texts, harder words, and more complicated sentences.

All Aboard for happy reading!

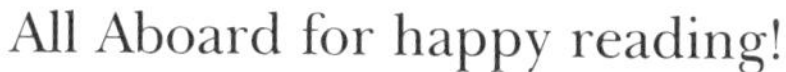

To Chris—J.D.

To Joan Farabee, Vickie Geckle,
and Gayle Reichert—cool teachers all—
and to Jackie, our black cat
(See if you can find him!)—H.P.

Special thanks to Brendon Hoch, the Lamont-Doherty Earth Observatory.

 Published by Grosset & Dunlap, Inc., a member of Penguin Putnam Books for Young Readers, New York. ALL ABOARD READING is a trademark of The Putnam & Grosset Group. GROSSET & DUNLAP is a trademark of Grosset & Dunlap, Inc. Published simultaneously in Canada. Printed in the U.S.A.

Library of Congress Cataloging-in-Publication Data

Dussling, Jennifer.
Pink snow and other weird weather / by Jennifer Dussling : illustrated by Heidi Petach
p. cm — (All aboard reading. Level 2)
1. Weather—Miscellanea—Juvenile literature. 2. Meteorology—Miscellanea—Juvenile literature. I. Petach, Heidi. II. Title. III. Series.
QC981.3.D88 1998
551.5—dc21
98-14336
CIP
AC

ISBN 0-448-41887-8 (GB) B C D E F G H I J
ISBN 0-448-41858-4 (pbk) C D E F G H I J

ALL ABOARD READING™

Level 2
Grades 1-3

Pink Snow and Other Weird Weather

By Jennifer Dussling
Illustrated by Heidi Petach

Grosset & Dunlap • New York

You are outside playing.

And it starts to snow.

Yippee!

But wait!

There is something strange

about this snow.

It is dark pink!

Pink snow?

Is that possible?

Yes!

Snow is not always white.

Every once in a while

snow can be a different color.

How does this happen?

Snow is made in clouds.

Sometimes strong winds

pick up tiny bits

of red soil and dust.

These bits of soil are blown

up into snow clouds.

Snow forms around red soil.

The snow looks dark pink!

Most people never see pink snow.

It is very rare.

It is very weird.

But sometimes the weather

does very weird things.

It is very hot

when it should be cold.

Or very cold

when it should be hot.

Or strange things rain down from the sky.

Here is what happened
one day in France in 1833.
Rain was falling on the streets
of a small town outside Paris.
People rushed from place to place
with their umbrellas.

Then all of a sudden,
something else started
to fall with the rain.
Toads.
Toads were falling from the sky!
They dropped on the rooftops.
They hit umbrellas.
Then they hopped around
in the wet streets!

The people of the town
must have been amazed.
And maybe they were afraid.
How did this happen?

Scientists think there is a simple answer.
Sometimes a special kind of storm
forms over an ocean or a lake.
It is called a waterspout.

The strong winds of a waterspout
whirl around and around.
A waterspout can suck up water.
It can suck up frogs or fish, too.

Sometimes the waterspout

will move over dry land.

When it starts to die out,

the frogs or fish fall to the ground.

In 1894, it rained jellyfish in England.

Other places have had
snails, worms, or even snakes
fall from the sky.

Like a waterspout,

a tornado is a storm with fierce winds

that whirl around.

Tornadoes are weird.

They move in crazy paths.

A tornado can crush one house
and leave the next one alone.

It can strip the bark off a tree
or pluck the feathers off a chicken.

In 1974, a tornado in Ohio

knocked down a farmhouse.

Everything inside was broken—

beds, chairs, tables.

Only three things were not broken.

A mirror, a case of eggs,

and a box of Christmas tree ornaments!

The unlucky town of Codell, Kansas,
is almost like a magnet for tornadoes.
A tornado hit Codell in 1916.
In 1917, a tornado hit Codell.
Again in 1918, a tornado hit Codell.

And here is the strangest thing.

The tornado struck each year on May 20—

the same exact day!

Some people say lightning never strikes
the same place twice.
That is not true.
Lightning hits the
Empire State Building
in New York City
about forty times a year.
So what, you say?
A building cannot get hurt
by lightning.
But did you know
one man was struck by lightning
seven times?

His name was Roy Sullivan,

and he was a park ranger.

One time he was fishing.

One time he was driving a truck.

One time he was in his front yard.

And one time he was even inside!

Lightning melted his watch.

It burned his hair.

But it didn't kill him.

Why was Roy Sullivan hit so many times?

Scientists don't know.

Lightning is just a bolt of electricity.

It can jump from a cloud

to the ground.

It can jump from the ground

to a cloud.

Lightning can even jump

from cloud to cloud.

Once a gas station worker
saw lightning hit a flock of pelicans
flying through the air.
It killed all twenty-seven of them!

That's just plain weird.

Here are some more

weird weather facts.

You may not believe them.

But they are all true!

In Montana in 1887,

the biggest snowflakes ever

fell from the sky.

Each one was fifteen inches across—

as big as a dinner plate!

In Hawaii,
there is one mountain
where it rains about 350 days a year!

Sometimes hard balls of ice

fall from storm clouds.

They are called hailstones.

Most hail is small and round.

But every once in a while,

a hailstone can be as big as an orange.

Or shaped like a star.

And one time a hailstone fell

with a turtle frozen inside!

Then there is the story of 1816.
The weather that year
was very, very weird.
In Europe and in parts of America,
1816 is known as
"The Year Without a Summer."

And it was all caused by a volcano.

It's true.

In April 1815,
a volcano erupted on an island
in the Pacific Ocean.
The volcano spewed
lots and lots of ash and dust
into the air.

People on nearby islands
did not see the sun for three whole days.
The ash and dust from the volcano
stayed in the air above the earth.
Then it drifted over other countries—
ones far away from the volcano.
It blocked out the heat from the sun.
It caused a cold spell.

Even a year later,

parts of New England

got six inches of snow...

in June!

There were bad frosts all summer long.

Crops died.

In Virginia,

Thomas Jefferson

had such a bad harvest on his farm,

he finally had to ask for a loan!

Most of the time

you don't even think about weather.

It is sunny or rainy.

Hot or cold.

But sometimes,

you can't help notice it!

So next time it rains,

watch out!

Who knows?

Maybe a frog will fall on your head!